Cambridge Elements

Elements of Paleontology

EQUITY, CULTURE, AND PLACE IN TEACHING PALEONTOLOGY

Student-Centered Pedagogy for Broadening Participation

Christy C. Visaggi
Georgia State University

CAMBRIDGE
UNIVERSITY PRESS

University Printing House, Cambridge CB2 8BS, United Kingdom

One Liberty Plaza, 20th Floor, New York, NY 10006, USA

477 Williamstown Road, Port Melbourne, VIC 3207, Australia

314–321, 3rd Floor, Plot 3, Splendor Forum, Jasola District Centre,
New Delhi – 110025, India

79 Anson Road, #06–04/06, Singapore 079906

Cambridge University Press is part of the University of Cambridge.

It furthers the University's mission by disseminating knowledge in the pursuit of education, learning, and research at the highest international levels of excellence.

www.cambridge.org
Information on this title: www.cambridge.org/9781108717939
DOI: 10.1017/9781108681766

First published 2020

A catalogue record for this publication is available from the British Library.

ISBN 978-1-108-71793-9 Paperback
ISSN 2517-7796 (print)
ISSN 2517-780X (electronic)

Equity, Culture, and Place in Teaching Paleontology

Student-Centered Pedagogy for Broadening Participation

Elements of Paleontology

DOI: 10.1017/9781108681766
First published online: July 2020

Christy C. Visaggi
Georgia State University

Author for correspondence: Christy C. Visaggi, cvisaggi@gsu.edu

Abstract: The diversity crisis in paleontology refers not to modern biota or the fossil record, but rather how our discipline lacks significant representation of individuals varying in race, ethnicity, and other aspects of identity. This Element is a call to action for broadening participation through improved classroom approaches as described in four sections. First, a brief review of the crisis and key concepts are presented. Next, culturally responsive pedagogy and related practices are introduced. Third, specific applications are offered for drawing cultural connections to studying the fossil record. Finally, recommendations including self-reflection are provided for fostering your own cultural competency. Our discipline offers much for understanding Earth history and contributing new knowledge to a world impacted by humans. However, we must first more effectively welcome, support, and inspire all students to embrace meaning and value in paleontology; it is critical for securing the future of our field.

Keywords: Diversity, Inclusion, Culturally Relevant Teaching, Culturally Responsive Pedagogy, Fossils & Earth History

ISBNs: 9781108717939 (PB), 9781108681766 (OC)
ISSNs: 2517-7796 (print), 2517-780X (electronic)

Contents

Introduction

The teaching of paleontology usually emphasizes an understanding of key concepts in the discipline (evolution, extinction), applications (biostratigraphy), and skills (fossil identification). Traditional lectures and labs have changed focus over the years as the discipline has expanded into new subfields. Technological advances have allowed for more complex approaches in both research and instruction. Growing knowledge of best practices in education has increasingly inspired the use of active learning strategies to better engage students in the classroom. However, as a discipline, we remain relatively unchanged, in that we are not doing enough to address the lack of diversity in our profession. This crisis can no longer be ignored.

The Crisis

The need for diversity in Science, Technology, Engineering, and Math (STEM) has gained attention over the last few decades, yet geosciences as a discipline remains alarmingly low for participation by underrepresented groups including women, people with disabilities, and racial/ethnic minorities (Huntoon & Lane, 2007; Velasco & Jaurrieta de Velasco, 2010; Stokes et al., 2014). Data evaluated by Bernard and Cooperdock (2018) from the NSF Survey of Earned Doctorates reported that between 1973 and 2012, while PhDs awarded to White women notably increased, numbers for minority groups overall have remained consistently low. Data from 2012 showed a breakdown of PhDs by race/ethnicity as 86% (White), 4% (Hispanic/Latino), 3% (Asian or Pacific Islander), 2% (Black), 2% (more than one race), 2% (unknown), and 1% (American Indian/AK Native). In the early 2000s, women represented only 16 percent of faculty in the geosciences (Holmes et al., 2008), and in the Paleontological Society, at the professional level (i.e., not including students), only 23 percent of the membership were women (Stigall, 2013). It is also painfully obvious year after year in attending conferences (e.g., Plotnick et al., 2014, on gender representation at the North American Paleontological Convention) that we are a homogenous group especially in regards to race/ethnicity and that our discipline may not be inclusive to all individuals. Volunteered data collected in 2017 as part of membership database updates for the Paleontological Society revealed similar demographic patterns of concern from the 20 percent response rate of members (Appendix A).

Furthermore, public perception of paleontology is muddled. Our discipline is often referred to as esoteric or merely futile “stamp collecting” with little or no value in addressing the challenges of our modern world. To be excited by the allure of dinosaurs is embraced in childhood, but paleontology has repeatedly been in danger of being deemed a science that is not worth investing in when government funding is scrutinized. We, as professionals, recognize how fossils have been instrumental in defining geologic time, correlating strata across distances, providing materials for economic resources (coal, natural gas, oil), as well as in examining evolution and how species respond to change, which is particularly relevant to the many issues we presently face as climate change and anthropogenic impacts alter our world. Yet, often the public is unaware of the utility of the fossil record in these important ways. I say all of this here because we, as paleontologists, need to better communicate our science to the masses, and that starts in the classroom with each and every student we encounter from the general introductory course to the graduate level. To effectively reach *all* students however, we must improve our pedagogical approaches, reflect on the nature and structure of our discipline, and ask whether we are being inclusive in our actions or whether we may be, in fact, unwelcoming. Thus, while it is essential to demonstrate the relevance of paleontology to all audiences (extending far beyond simply educating the next generation of paleontologists), we must also prioritize broadening participation in our discipline as well. Having professionals that represent a range of abilities, backgrounds, and identities is critical for a variety of reasons.

Research has shown that diversity is essential for bringing new perspectives to scientific problems (Wilson, 1992; Medin & Lee, 2012; Guterl, 2014; Phillips, 2014; Smith-Doerr et al., 2017; Powell, 2018), and in order to promote creativity and progress, we must work harder to lift up groups that are traditionally underrepresented in our field. Furthermore, diversity is the future. The US Census Bureau projected that by 2020 more than 50 percent of children are expected to belong to a minority racial or ethnic group as part of a so-called ‘browning of America’ (e.g., Chávez & Longerbeam, 2016), and as recently as 2015, a shift is already apparent in the next generation, in that babies of color now make up more than half of all new births in our nation. Slowly, we are seeing this shift in the makeup of college students, making it imperative to ask whether our existing pedagogy is effective in diverse classrooms. Even if an institution is predominantly composed of a particular racial group or socioeconomic class, mindfulness in our approaches to be more inclusive is critical in strengthening the importance and place that diversity has in science and in the public view. This need is not merely a “goal” per se, but as with all work in broadening participation, making it a *priority* is essential (Inclusive Astronomy, 2015). We must realize as educators, that who we are as individuals matters in all aspects of our lives, and in guiding the future of our discipline, we must work first to understand barriers that contribute to the homogenization of our field, and then consider how to improve things as practitioners in the classroom.

Obstacles

Why does the lack of diversity exist in the geosciences and related STEM fields? There are many contributing factors (see Inclusive Astronomy, 2015, for an in depth review of common barriers), and while we may not be able to address all barriers in the classroom, acknowledging the ways in which monocultural views of our field impact our ability to broaden participation is an important first step. The history of our discipline, how it is perceived, and how it is portrayed in popular media has led to the characterization that paleontologists are often composed of rugged (White) males who work in remote field locations dusting off dinosaur skeletons under intense conditions. This depiction does not reflect the spectrum of what paleontologists do or who we are, although admittedly, able-bodied, cis-gender White males have dominated the discipline especially in how the field is presented from a Eurocentric view; that is, in how the history of paleontology has largely focused on Western civilization. There are females and members of under-represented groups who have contributed to paleontology, but until recently (e.g., Gold & West, 2017; Stricker, 2017; Bearded Lady Project, 2018), very little attention was given to these individuals aside from Mary Anning of England. As a result, even if students belonging to minority groups are interested in fossils, it may be that they are unable to visualize paleontology as a viable career option. Such students may not feel as though they "belong," and might fall victim to inherent biases such as 'stereotype threat' or 'imposter syndrome' as has been reported across STEM fields (Steele, 1997; Gonzales et al., 2002; Huntoon et al., 2015). STEM identity is a leading factor known to contribute to retention issues for underrepresented groups (e.g., LGBTQ students, Hughes, 2018), and this problem is indeed pervasive in geosciences (Black, 2019; Goldberg, 2019). Students from diverse backgrounds also may not have gained the necessary academic preparation to succeed in pursuing careers in STEM (Seymour & Hewitt, 1997) and/or may lack support or role models who might otherwise encourage their interests (O'Connell & Holmes, 2011, as reported for STEM). Public perception plays into this issue as well, in that first-generation students and people of color or of less economic means may have family members who do not actively promote an interest in paleontology as it may seem a frivolous or unsteady choice for employment compared to high-paying marketable jobs in medicine that are likely more promising or respected. Family criticism and fewer informal experiences in the outdoors have been cited as additional factors that may limit students of color from studying geosciences compared to their White peers (e.g., Stokes et al., 2014). In addition, students with less privilege may be attending institutions that lack resources or opportunities to study the fossil record, which is why programs that specifically seek to expand access to authentic research such as the National Science Foundation Research Experiences for Undergraduates (REU) are so important, as students from across the nation, including at community colleges, can apply. Geosciences suffers even further due to the fact that Earth science is often not well represented at the K-12 level, leading to less exposure before students arrive at college, thereby limiting the potential for students to enroll in courses where they might learn about the fossil record in depth. See Stokes et al. (2015), Sherman-Morris and McNeal (2016), and Sexton et al. (2018) for additional information regarding perceptions that students have at the undergraduate level which factor into selecting geosciences as a potential major including differences related to gender and race/ethnicity. Furthermore, even if students initially show an interest in pursuing coursework or career paths in geosciences, factors such as racism, including that imposed by individuals (explicit and implicit), institutions, and society continue to contribute to the lack of diversity in the geosciences overall (Dutt, 2020).

While certain barriers mentioned here cannot be solved single-handedly by altering pedagogy in the classroom, what we can do is be mindful of these issues and consider how better we can address specific aspects of these challenges through student-centered instruction that is more likely to be welcoming to individuals from a variety of backgrounds and promote their success moving forward. Such culturally informed approaches are necessary regardless of the composition of the audience in a classroom, otherwise we continue to carry on the perception that paleontology is a discipline reserved primarily for privileged, able-bodied White males who want to work in the field on dinosaur bones. Bringing in diverse perspectives, emphasizing cultural connections, utilizing constructivism in building on any pre-existing knowledge of students, and fostering a sense of unity through collaborative work, are all steps in the right direction. Furthermore, we cannot rely on existing professionals from diverse backgrounds to carry the labor needed to serve as role models and advocate for underserved groups (Jimenez et al., 2019). We must all learn how to be more inclusive and promote diversity in our classroom if we are to overcome this crisis in our discipline. Understanding the inequities that exist and committing to a change in practice is essential in order to address the disparities that continue to affect the experiences of diverse students who have been underserved in their education (Williams & Lemons-Smith, 2009). This is critical not only as a social justice issue, in that all individuals should have opportunities and support to pursue any career path that they choose, but it is also vital for progress in scientific fields including paleontology.

A Call for Action

Teaching practices labeled as "culturally relevant" originated out of a desire to address achievement gaps of students who were not succeeding in US public schools, particularly individuals of color, lower Socioeconomic Status (SES), and English Language Learner (ELL) speakers. Ladson-Billings (1995, 2014) characterized the pedagogical approach as centered on principles of academic success, cultural competence, and sociopolitical consciousness. She acknowledged research by anthropologists in the 1980s who examined "culturally congruent," "culturally appropriate," "culturally responsive," and "culturally compatible" ways to strengthen connections between school and the home culture of a student to improve academic success. Her work initiated a movement of educational reform, and as stated in her culturally relevant pedagogy 2.0 'remix' paper, "Instead of asking what was wrong with African American learners, I dared to ask what was right with these students and what happened in the classrooms of teachers who seemed to experience pedagogical success with them" (Ladson-Billings, 2014, p. 74). Through that lens, we must examine the variety of approaches that integrate aspects of culture into a pedagogical framework for learning, as the research is clear: meaningful cultural connections and considerations of how people learn as related to culture can be effectively used to enhance the success of underserved students in the classroom (Irvin & Darling, 2005; Aronson & Laughter, 2016; Byrd, 2016).

Diversity, Inclusion, Equity, Culture: Definitions

The purpose of this Element is to highlight the homogenous nature of our discipline, acknowledge barriers to the success of underrepresented students, and put forth a call to action. The subsequent sections of the Element are divided into an emphasis on cultural competency in approaching pedagogy and expanding access to learners through inclusion and improved assessment practices, considerations for specific applications to make instruction of paleontology more meaningful through cultural contexts, followed by an emphasis on self-reflection prior to the final recommendations. However, before proceeding further, I offer a brief orientation into a few key concepts and language pertinent to discussions of pedagogy focused on diversity and inclusion in striving for equity in the classroom.

Diversity is often used as a way to specifically reference students who are diverse linguistically, culturally, and in regards to race and ethnicity (e.g., Adams et al., 2017). However, it can encompass many more aspects of a person's identity. Furthermore, it is critical to acknowledge that identities are intersectional (e.g., a male student may be cis-gender, straight, unmarried, neurodiverse, able-bodied, hearing-impaired, learning-disabled, Black, middle-class, Jewish, highly educated, Hispanic, and uses pronouns they/them/theirs). It is important not to make assumptions or generalize students based on simplistic or narrow views of identities.

Inclusion focuses on ensuring that all individuals are valued in how classroom learning is constructed, especially students who might otherwise feel marginalized or excluded for any number of reasons. Typically, such practices involve strategies designed to remove impediments to learning in an effort to better meet the unique needs of all students. Classroom approaches are intended to better support and encourage participation by all learners such that students who vary in abilities, backgrounds, and identities feel as though they belong and can effectively engage in ways that promote their success in a community of learners.

Equity emphasizes providing learning opportunities and resources in support of all individuals in a classroom by recognizing that the "same" does not mean "equal" in regard to access, interests, and the needs of all individuals. Students might be disadvantaged coming into the classroom in knowledge or skills, could be resource-limited outside of school, vary in their ability and how to learn due to interpersonal, cognitive, or physical challenges, or perhaps unable to catch up or otherwise get ahead in their education for a variety of reasons. It may be the case that students of color or of lower socioeconomic status are further behind compared to their peers as a result of less access and opportunity prior to college, which is why it is critical to recognize students as individuals in order to address any disparities that may limit the success of all learners. To achieve more equitable classrooms that support a diversity of learners, approaches that are welcoming and inclusive of all individuals are essential.

Culture, as referenced in this Element, focuses primarily on the constructs of race and ethnicity and any related contexts such as language, beliefs, values, and norms. The purpose of this approach is to highlight culture as a means by which we can broaden participation of underrepresented groups in paleontology via improved cultural competency in our pedagogy (although characterizations of culture that are not explicitly linked to racial/ethnic groups are additionally included). Pedagogy that more specifically highlights accessibility and inclusion, particularly with respect to expanding diversity to include other forms of identity such as disabilities or how learners may vary in skill or processing and communicating knowledge, are also briefly woven into the discussions below.

Pedagogy of Culture and Inclusion

Theory and Practice

Teaching practices that incorporate culture are defined differently depending on how elements of culture are used in structuring pedagogy. Terms in the literature that overlap conceptually in advocating for cultural competency as essential to education but are not quite equivalent in meaning include, although are not limited to, the following: culturally relevant (Ladson-Billings, 1995), culturally responsive (Wlodkowski & Ginsberg, 1995; Gay, 2000; Hammond, 2015), culturally strengths-based (Chávez & Longerbeam, 2016), culturally-revitalizing (McCartney & Lee, 2014), and culturally-sustaining (Paris, 2012; Paris & Alim, 2014). Even if authors use the same suite of words to label their pedagogy, that may not mean they are referencing parallel classroom approaches (see Brown-Jeffy & Cooper, 2011, and Aronson & Laughter, 2016, for a synthesis of research and practice that further explores several of the models referenced above). Specific practices highlighted in this Element emphasize different aspects of how culture can be considered as a foundation for improving student outcomes following recommendations described by Wlodkowski and Ginsberg (1995), Hammond (2015), and Gay (2018). Through this window into similar, yet different, visions of how culture can be used as an educational framework (as well as a note of caution below to avoid misinterpretations before practices are implemented), we can consider how best to effect change in paleontology upon desiring to be more equitable and inclusive to enhance the participation of underrepresented groups in our discipline.

Misconceptions

There are several misconceptions that many authors caution against in reflecting on how to adopt new classroom strategies that use culture as a means for broadening participation and promoting the success of diverse students that should be acknowledged before exploring how to proceed in practice. For example, if seeking to attract more Black students to the discipline, simply adding examples of paleontological discoveries in Africa that may be more appealing to that audience is not necessarily appropriate in isolation. Even though cultural relevance is desirable and should be incorporated, it is not enough to attempt to insert culture into learning but rather that learning should be offered within the context of culture (Pewewardy, 1993). Approaches that are *culturally responsive* go well beyond the use of examples that might appeal to certain audiences by considering *how* best people learn. In addition, simplistic interpretations of infusing culture into lessons can be problematic thereby requiring a paradigm shift in how culture is utilized in pedagogy (Sleeter, 2011, 2012; Ladson-Billings, 2014). Furthermore, race and culture are often confounded as being interchangeable in meaning, which can lead to problems in the implementation of approaches meant to support diversity (Irvin & Darling, 2005). Culturally responsive pedagogy is not equivalent to multicultural education even though it may be misconstrued as being the same (Hammond, 2015).

The celebration of diversity that is often a focus of multicultural education does not incorporate pedagogy that is based on how people learn as a reflection of their identities. Ideas such as collectivism in learning, importance of building learning capacity, bridging connections, addressing implicit bias, and providing affirmation and validation for students are components of a view of culturally responsive pedagogy that may not even require an explicit incorporation of race/ethnicity, and such methodologies can be successful in supporting all students regardless of their specific backgrounds (Hammond, 2015). To that end, approaches characterized as culturally responsive can be even more inclusive than are originally perceived to be in name, in that all learners gain from such efforts in considering culture as more than simply elements related to race or ethnicity (e.g., gender identity, sexual orientation, neurotype, learning/physical disabilities, class, urban vs. rural, etc.). Finally, in seeking to embrace culture in broadening participation in the classroom through pedagogy, caution is needed in order to avoid instances of appropriation, that is, even if the intent is seemingly positive, it is essential to avoid inappropriate emulation of culture and the related negative impacts that could result in such misguided attempts to strive for cultural competency in pedagogy.

Teaching Practices

There is no single model that is universally accepted regarding how best culture can be utilized as a framework in striving for the academic success of diverse and all students in the classroom. However, commonalities in various pedagogical approaches include a respect for the cultural identities of all learners, drawing on the existing knowledge of students, and encouraging any intrinsic motivation for learning as influenced by emotions and cultural beliefs instead of focusing on institutional rewards (Wlodkowski & Ginsberg, 1995; Gay, 2018). Furthermore, pedagogy need not necessarily 'look' cultural as such approaches can be useful for all students from a neuroscience perspective (Hammond, 2015). For example, incorporating specifics regarding a particular underrepresented group should not be the only focus, but rather instruction that fosters the active engagement of how all students can best participate, learn, and communicate as a community is what is needed. Approaches that are student-centered with collaborative learning that is active, inquiry-based, experiential, or focused on projects or problem-solving can serve as a foundation for better supporting all students in an inclusive manner and for engaging diverse learners in the absence of specific cultural references. Likewise, it is important to consider our own identities and assumptions in how we, as instructors, learn about the world as that impacts how we approach guiding our students in the classroom (Hailu et al., 2017).

Table 1 Themes that encompass varying perspectives of how culturally responsive pedagogy can be implemented and recommended approaches

Students as Individuals	Engaging Practices	Bridging Connections
Learn all names and demonstrate support for student-centered learning	Promote ***small groups learning*** by setting up collaboration opportunities	Bring in ***guest speakers and inspiring role models*** who are best suited to relate to students
Endeavor to know ***students as people*** and model a welcoming atmosphere for all individuals	***Gamify lessons*** and capitalize on the energy and fun that brings to learning	Make use of ***relevant quotes, vocabulary, and metaphors*** in being mindful of all audiences
Provide opportunities for differentiated instruction such as via use of ***learning stations***	***Tell stories*** and have students create narratives to demonstrate understanding	Integrate ***appropriate and positive media*** in desiring to reach and empower all students
Encourage ***ownership*** and contribution of student ideas in assignments and projects	Explore ***real world problems*** that can be motivating in drawing upon a need to learn	Call upon the sense of place by using examples of ***familiar and meaningful places*** to students
Consider the range of ***access and opportunities*** that students have coming into a classroom	Develop opportunities for learning experiences that ***build community***	Integrate learning about subject material via ***cultural contexts*** by highlighting such connections

To effectively embrace and strive for a mindset as instructors that is centered in culturally responsive pedagogy, I offer here a framework that highlights how better to support diversity and inclusion in the classroom by a) recognizing students as individuals, b) utilizing engaging practices for learning, and c) bridging meaningful connections. Table 1 highlights a collection of approaches that fall under the umbrella of what could be considered as culturally responsive and inclusive depending on the model, whereby students are valued as individuals who vary in needs and interests, activities that foster active participation and community learning are emphasized, and specific connections as related to culture or identity allow students to identify with class material in more meaningful ways. Each of these categories is discussed in more depth below and accompanied by examples that I utilize in my own classroom.

Students as Individuals: Motivation for learning is influenced by personal values and goals. Allowing students to explore and build upon their own learning experiences provides an opportunity to foster the academic growth of individuals in the classroom (e.g., Hailu et al., 2017). Giving students options in claiming ownership over what they learn makes educational experiences more meaningful with lasting impacts. For example, in my paleontology course, I have purposefully given choices to students to make learning more appealing by allowing them to select papers of their choosing in doing literature critiques, selecting a particular mass extinction or major event in Earth history to study in depth from multiple perspectives (biological, chemical, geological), or utilizing any variety of everyday objects to demonstrate an understanding of phylogeny, etc. Likewise, I have set up stations that allow students to explore concepts in paleontology across multiple modes of learning. In addition, students might be given the chance to individually study specimens at their own pace or navigate learning in flexible groups that vary depending on the learning objectives and the unique skills that individual members may bring to group work. Providing an opportunity for students to answer open-ended questions allows for beginner through advanced interpretations that can both accommodate and challenge the needs of all learners as well as promote an opportunity for individuals to more deeply connect to the material based on how they learn and what is meaningful to them. See Santamaria (2009) for more on differentiated instruction as related to culturally responsive pedagogy.

Engaging Practices: The immensity of the fossil record in understanding the sequence of events through geologic time can be daunting for students to "memorize" when presented to them. To make it more engaging and spark curiosity in learners, I have changed my approaches such that in any classes in which students learn about the history of life and Earth, instead of reading or listening to me, I have them begin by studying a set of twenty-four picture cards for different events in Earth history. They work in pairs discussing their reasoning for what they think the order of events might be as they work to arrange the cards in order from the formation of Earth to the present day. One of my students came up with the idea for this activity in order to 'make it social' and 'gamify' the lesson to enhance collaborative active learning of Earth history.

Over the years, I have worked to incorporate more storytelling as a way to communicate science and articulate learning outcomes as well. Traditions of oral storytelling exemplify how learning is embedded in culture (Hammond, 2015); similarly, humans have a way of using stories in order to make sense of events in our lives (Gay, 2018). Educators could more frequently organize ideas and experiences into meaningful accounts of ordered happenings in instruction to engage all students just as individuals do in sharing personal experiences. ArcGIS story maps are an example of how I have had students "tell stories" about what they have learned regarding anthropogenic impacts and changing reefs through history in comparing ancient and modern marine environments in the Bahamas.

Bridging Connections: Offering materials from diverse authors, cultures, and identity perspectives can demonstrate valuable insight to the discipline that helps to eliminate prejudice and perceptions of inferiority against people with disabilities, students of color, international students, ELL students, etc. Given goals of broadening participation by fostering a more inclusive classroom for learning, elevating cultural connections in classroom approaches can demonstrate how diverse knowledge systems are valued and related to learning (Hailu et al., 2017). Traditional Ecological Knowledge (TEK) is gaining attention in addressing global conservation and sustainability issues as Indigenous peoples have acquired immense knowledge of the natural world in living off the land and sea for generations (e.g., Senos et al., 2006). Thus, learning about changes in Earth history and modern shifting baselines can be linked to experiences and perspectives of such underrepresented groups providing an opportunity to value insight from non-dominant cultures that can be incorporated into the paleontology classroom. Other meaningful connections in learning about the fossil record could draw on the use of native languages (Meyer & Crawford, 2011; Meyer et al., 2012), injustice related to fossil discoveries (Mayor, 2007a), significance or use of fossils in different cultures (Duffin, 2008), and related place-based connections (Mayor, 2007b; Hughes et al., 2015). These examples provide a glimpse into how meaningful connections can be related to learning paleontology; specific applications are explored in more depth later in this Element upon considering how to approach implementation.

Community and Inclusion

To emphasize a need for inclusion as part of broadening participation and fostering a community of learners that incorporates but moves beyond the primary cultural and identity focus of this Element, additional considerations for pedagogy should be examined such as Universal Design for Learning (UDL) – an approach that is increasingly being used to enhance equity and access in education (CAST, 2018). The principles of UDL are based in providing students multiple means of representation, engagement, and expression in an effort to better serve the differences that make up an array of learners in the classroom. For example, students may have learning or sensory disabilities (or vary in language or cultural backgrounds), all of which can uniquely influence their learning process. There is no single approach that works equally for all, so it is important that options exist for learning that range in how material is presented (e.g., audio/visual). The use of virtual reality might be a way for more students to be included in learning opportunities that were previously restricted to only those students who could handle extreme physical exertion to access localities. Similarly, examples of microfossils that are scaled up using 3-D printing might be interesting to all students but particularly beneficial for any with impaired vision. Driving factors in motivation and how students best engage in learning also vary by individual. Novelty, routine, group/independent work, intrinsic factors related to their identity, and more can influence how receptive a student may be in learning; differentiated instruction can help in meeting the needs and interests for each and every learner (Patterson et al., 2009). How best individuals navigate an educational experience, and can demonstrate what they know, differs by person as well. Students may be strong in speech but not writing or may be impacted by limited mobility or executive function disorders; offering multiple means for communicating knowledge is ideal. For example, students who are new to the English language might fare better in being given a chance to verbally express their knowledge rather than struggling to find the words to explain what they know in writing. Conversely, students who are introverted or struggle with social skills, may do better in written exams.

The inclusive classroom with respect to disabilities is not the main focus of this Element. However, nearly 1 in 4 people have a disability according to the US Census Bureau, and yet students with disabilities are not often acknowledged in efforts to broaden participation from underrepresented groups. The paleontological community lacks such representation and it is well-documented that individuals with disabilities are underrepresented in the geosciences broadly (Atchison & Martinez-Frias, 2012; Carabajal et al., 2017). More intentional instructional approaches are needed to improve access and accommodations for students that require a more inclusive classroom in order to broaden their participation in the discipline, whether they are mobility-impaired, sensory-impaired, etc. (DePaor et al., 2017; Goring et al., 2017). Building community among learners is a critical step to embrace underserved groups in the classroom; lessons regarding inclusion are important for all given the goals of culturally responsive pedagogy. Furthermore, because field experiences are often utilized in learning paleontology, special attention should be given to inclusive instruction in the field as many underrepresented groups have faced challenges in that setting. To learn more about these issues as well as to access helpful resources, refer to the "In the Field" section of the Advance GEO partnership website hosted by the Science Education Resource Center (SERC).

Table 2 Practices related to pedagogy from the perspective of assessment to consider in fostering a more equitable, inclusive, and culturally responsive experience for all students

Do	Don't
Offer flexible deadlines for all individuals as under-resourced students may have more challenges in accessing resources limiting their ability to meet deadlines	Place demographic questions at the start of an exam as ***it can reinforce stereotypes*** impacting success
Consider your own biases as an instructor in assessing student work, what is fair to expect of students, and what is influenced by culture	Exclude identities in soliciting demographic data in any sort of survey used as that ***can be isolating*** (e.g., only single races, binary choices for gender)
Reinforce that academic ability is not fixed and instead ***foster a growth mindset*** in students	Rely on standardized tests as they are a ***poor predictor of student potential*** for marginalized and oppressed groups

Equity and Assessment

Teaching is only part of the pedagogy equation in reflecting on how best to support students in a classroom of varying identities, needs, and abilities. Approaches to assessment of learning outcomes may require modification as well in striving to be inclusive and culturally responsive while additionally addressing the achievement (or rather, opportunity gap) noted by numerous authors. This Element is not meant to cover assessment at length, but several considerations as related to instruction are offered in Table 2. Finally, in striving to foster a classroom that is intentional in desiring to improve equity and broaden participation via culturally responsive and inclusive practices, adding a statement on your syllabus that starts the conversation with students helps to demonstrate how you value diversity as an educator and that you seek to improve the learning experience for all individuals in a class (Appendix B).

Applications in Paleontology

Utilizing connections in the classroom that are meaningful to students is important regardless of any specific suite of culturally responsive approaches used to foster a strong community of diverse learners. There are many ways in which cultural relevance could be considered in reframing a class focused on paleontology or Earth history. Three different perspectives are offered here in contemplating how connections can be explored as related to the a) audience (demographics of students), b) concepts in the discipline, and c) places of significance to "cultures" broadly defined.

Audience

Thinking about the composition of individuals in a classroom is essential for knowing how best to guide student learning as an instructor. Recognizing your own assumptions about the classroom audience and working to better understand the knowledge, skills, and attitudes students have coming into a course is critical as that information can be leveraged to enhance their engagement and learning. Pedagogical approaches that promote diversity and inclusion should be used regardless of the audience; however, choice of particular instructional approaches may be influenced by classroom demographics in striving to effectively reach all students in achieving learning outcomes. The goal of the May 2017 Teaching Geosciences with Pan-African Approaches workshop organized by the Science Education Resource Center (SERC) had this focus in mind. Participants collaborated on ways to engage students in learning about geosciences through the lens of Pan-Africanism as part of efforts to broaden participation of African and African American students in the classroom and discipline overall. For example, rhythm is a principle of Pan-Africanism (Hewitt, 2017). Topics reflective of "rhythm" are abundant in what we can learn from the fossil record (e.g., cycles of changes in sea level, climate, and supercontinents). Sankofa is another tenet of Pan-Africanism coming from Ghana, and in relation to the African Diaspora has come to mean that one must look backward in order to successfully move forward. Conservation paleobiology stresses the importance of paleontological perspectives in addressing modern conservation problems (and in doing so offers a connection to introduce the concept of Sankofa, in that the fossil record provides a window into the past necessary for future conservation). Conversations regarding geological resources or paleontological discoveries that may be particularly meaningful to students of African ancestry were additionally discussed as part of the workshop such as the state fossil of South Carolina, a woolly mammoth, because the history of discovery was that a fossil tooth was originally found by an enslaved person on a plantation in 1725. Note here that use of the word "slave" was not chosen to describe the individual who discovered the fossil but rather the adjective "enslaved" as this more appropriately acknowledges that slavery was in fact imposed on the identity of people and thus offers a more culturally-informed understanding of the devastation people of color have faced in human history when sharing this story.

Topics

Concepts in paleontology can be easily presented in the classroom alongside cultural connections that may offer more meaning to students in studying the fossil record as part of an overall shift in pedagogical approaches designed to reach all learners. For example, understanding factors that contribute to extinctions in the fossil record is a key element of any class that examines the history of life. The modern "sixth" extinction is an obvious choice for demonstrating the relevance and value of paleontology in our present world and the role that human activities have played since the Pleistocene. Yet, how often do we dive into the cultural context of species at risk or consider the many ways in which aspects of culture may be deeply connected to the livelihood of various species? Several publications in recent years have focused on the biodiversity crisis of the Anthropocene by examining the cultural significance of species, viewing conservation measures through a cultural lens, and how humans may perceive or experience their actions as part of this ongoing extinction (e.g., Sodikoff, 2012; Heise, 2016). In addition, it is helpful to recognize that many Indigenous cultures have a strong history of awareness in how human activities can disrupt the delicate balance of the natural world as such communities have relied heavily on these ecosystems. More examples of how concepts in paleontology can be connected to culture are provided in Table 3.

It should be noted that the examples featured in this section may be more relevant to particular audiences or concepts in paleontology but should not be viewed as a checklist for achieving cultural competency. The inclusive classroom that welcomes diversity is more than an infusion of culturally relevant material in lessons but a complete shift in pedagogy overall. However, it is my hope that by bringing attention to specific examples and resources, more educators can consider how aspects of culture can be integrated into the paleontology classroom as a first step in reframing pedagogy to be more culturally responsive and decolonizing the narrative (e.g., Yusoff, 2019). Connections between fossils and culture have been a part of human history from ancient civilizations to the present day and offer an interesting link to enhancing meaningful instruction of paleontology. Specific resources not yet mentioned that could be useful in exploring additional cultural connections are listed in Table 4.

Table 3 Themes in paleontology that can be viewed through a cultural lens by reflecting on how concepts may have connections or meaningful significance to students of diverse backgrounds

Themes	Reflections	Examples
Evolution	Being sensitive to family dynamics and staying away from reconstructing family lineages as information may be missing or stir up personal issues that could inhibit learning.	Traditions that are passed down over many generations in a broader cultural context could be good choices for discussing evolutionary concepts in a way that is familiar and meaningful.
Taphonomy	Burial is an important step in processes of fossilization and is a practice humans have engaged in for over 100,000 years with unique customs. How humans bury and honor the dead has deep cultural connections that could be integrated into learning about paleontology if done respectfully and does not lead to cultural appropriation.	Día de los Muertos (Day of the Dead) is a holiday that originated in Mexico and is now widely celebrated among Latinx populations. Learning about skulls such as in studying hominid evolution provides an opportunity to honor the rich cultural connections that this holiday embodies as identified through symbolic skeletons (calacas) of this day.
Paleoclimate	Changing climates have altered the course of human history since the very beginning of our species in Africa. More examples include the migration of Native Americans over land bridges during the last Ice Age into the Americas and how modern warming and sea level rise are severely impacting communities worldwide.	Ethnographies of how humans and cultures are impacted by climate change, especially low-lying coastal communities such as the struggling Iñupiaq Eskimos in Alaska (Marino, 2015) could be brought into class discussions when examining the impacts of modern climate changes and the value of paleoclimate reconstructions.

Paleoecology & Paleoenvironments	The utility of the fossil record in studying our modern world via conservation paleobiology offers a window for examining natural and human-induced changes to landscapes, food webs, and resource utilization.	Insight from paleoecological analyses is increasingly being used in building an understanding of changes over the last few million years including remote locations with Indigenous populations (e.g., Easter Island: Rull & Giralt, 2018).
Mollusca	Mollusks have a long evolutionary history and are geographically widespread. They are utilized in studies of predation, paleoclimate, and more. They remain economically important in the modern world and have significance in cultures around the globe as a source of food, use of shells in jewelry, etc.	Ex: *Mercenaria mercenaria* (quahog) is a common clam in the northeastern US. The beautiful purple and white shell material from along the inner margin has had a long history in being used in the making of wampum by Native Americans of the Eastern Woodlands for ornamental purposes (and only briefly had use as a currency due to European settlers).
Pollen & Plants	The study of paleoethnobotany (relationships between people and plants) offers a unique lens to learn about plants as related to culture. Plants have been used for food, paint, decoration, medicine, clothing, building material, and much more.	The rise of the grasses in the Cenozoic can be linked to ways in which humans make use of grasses across cultures. Ex: Basket weaving of sweetgrass is an art passed down through generations among the Gullah community of the coastal South with origins in West Africa

Table 4 Example resources that can be used to examine additional connections between paleontology and culture

Topics	Authors
Dinosaur excavations in Africa	Jacobs (1993)
Mythology and history of paleontology	Mayor (2000)
Fossil legends from Native Americans	Mayor (2005)
National Park Service: paleontological specimens and cultural contexts • Pueblos built from Triassic petrified wood in Arizona • Shark teeth used in jewelry by Hopewell Indians in Ohio • Legends of mammal fossils from the Lakota Sioux in Nebraska • John Wesley Powell collections during siege of Vicksburg, MS	Kenworthy and Santucci (2006)
Human anatomy, early vertebrate ancestors, discovery of *Tiktaalik*	Shubin (2008)
Got lactase? The co-evolution of genes and culture (film)	HHMI Biointeractive Video (2013)

Place

To strengthen student interest and knowledge in the classroom for all learners, a specific approach that is rapidly gaining ground in geosciences with benefits to enhancing diversity is place-based learning. This particular approach capitalizes on meaningful places of interest to students and by doing so builds upon existing knowledge they may have about their surroundings and any emotional attachments to such locations (sense of place). It is inquiry-driven and can lead to an improved understanding of concepts across disciplines through a lens centered in place. Project-based or experiential learning is often incorporated as well as opportunities for students to learn in places beyond the classroom. Ties between place-based methods and culturally responsive pedagogy are numerous and can be demonstrated using of a variety of approaches (Gruenewald & Smith, 2014). Research on place-based education has revealed that benefits can include enhanced curiosity and academic success, stronger ownership of educational goals, a heightened appreciation for such places and increased retention of under-represented groups in STEM including in geosciences (e.g., Sobel, 2004; Semken & Butler Freeman, 2008; Semken et al., 2009). Techniques in place-based education can draw upon cultural contexts that reflect race/ethnicity (Semken, 2005), but other populations of students may be connected to places for different reasons such as living in urban or coastal settings that can be used for studying geosciences (e.g., Davies, 2006; Kirkby, 2014). Being mindful of whether students are united by being in the South, rural mountains of Appalachia, along the active margin of the Western US, or converge by attending a remote small college in the Midwest or New England, becoming familiar with places of meaning to students at home, school, due to ancestry, or otherwise can be a useful framework for maximizing student learning in ways that align with culturally responsive pedagogy. Assessment practices can be better informed by considering reflections of culture and place in pedagogy as well enhancing the success of diverse learners (Geraghty Ward et al., 2014).

Five essential characteristics for place-based geoscience education are outlined by Semken (2005) as 1) learning is done through the lens of place, 2) diverse perspectives of that place are acknowledged, 3) instruction includes learning in that place or via materials derived from that place, 4) sustainable living in that place both ecologically and culturally is promoted, and 5) sense of place for the instructors and students is enriched. Techniques that fall under the umbrella of being culturally responsive then are often mirrored in place-based methods, in that they are both based in constructivism, consider personal experiences and identities, promote an active and collaborative learning community, draw upon emotions in bridging meaningful connections to learning, and can be particularly helpful for engaging students who belong to under-represented groups. Because students at Georgia State University (GSU), a minority-serving institution in the metro Atlanta region where I work, overwhelmingly come from the 159 counties in Georgia, use of local landscapes as a foundation for studying fossils and the history of life offer an intersection for learning in ways that leverage the benefits of both culturally-responsive pedagogy and place-based education. While many paleontology courses no doubt draw upon local landscapes for studying fossils in the field, learning about the fossil record though the lens of place drives all aspects of my course design at GSU as highlighted in Table 5. Use of place-based methods does not necessarily mean that "classic" examples beyond local contexts are ignored or not used, but rather that instead of simply defaulting to covering such examples, familiar and meaningful landscapes are elevated as a focus for learning material and integrated into all aspects of the course framework. This structure offers an opportunity to study paleontology in a way that builds on local knowledge and interest

Table 5 The class framework for Principles of Paleontology (GEOL/BIOL 4011) at GSU. Examples are highlighted for all course components that show connections to local and regional landscapes by featuring specimens and resources used in addressing focused content areas. Place-based approaches provide the foundation for learning throughout all aspects of the course.

	Focus	**Specimens**	**Resources**
Class Activities	Comparison of fossil-rich slabs from different units in the state for paleoenvironmental interpretation as studied by physiographic region	Ex: Ordovician limestones (brachiopods) and Mississippian chert (crinoids) from the Valley & Ridge, Pennsylvanian shale (plants) from the Appalachian Plateau, Oligocene limestone (echinoids) from the Coastal Plain	Roadside Geology of Georgia (Gore & Witherspoon, 2013)
Labs	Studying aspects of phyla (morphology, mineralogy, paleoecology) with an emphasis on examples from Georgia	Agnostids in Cambrian shales from Northwest Georgia, dentition of mastodons vs. Columbian mammoths from the Coastal Plain	Giant 16' x 20' floor map of the state of Georgia from National Geographic
Exams	Use of specimens on the lab practical as related to field excursions Short answer questions that require utilizing examples from local landscapes	"Describe form genera and their significance with reference to plants found in Northwest Georgia" "Describe the importance and geographic context of *Georgiacetus* in the evolution of cetaceans in the Cenozoic"	PaleoPortal, Publications of the Georgia Geological Survey

Field	Identifying fossils, describing sections, making interpretations of paleontological assemblages	*Archimedes* in Mississippian limestones Comparing preservation of Mollusca vs. Echinodermata in Cenozoic deposits of the Coastal Plain	Rockd, local museum collections, abandoned and active quarries
Assignments	Literature discussions that contextualize the class research project from a biogeographic perspective through reading, writing, and study of maps	Examining stratigraphy and existing data on fossil deposits recorded in the region Using class prompts to generate reflections on the regional significance of literature read in class given the focused research project	Macrostrat, Paleobiology Database
Research	Examining the paleoecology of marine faunas from stratigraphic units of the Plio-Pleistocene across extinction episodes	Processing bulk samples of molluscan specimens from the southeastern US that students sieve, sort, identify, analyze, and in the following semester report on at regional conferences	Neogene Marine Tropical Biota of North America, Neogene Atlas of Ancient Life: Southeastern US

in surroundings that students may already have from living in Georgia, including having students gain authentic research experience by studying fossils in the region as an overarching goal in the course. This approach, known as a Course-Based Undergraduate Research Experience (CURE), can be particularly beneficial for underserved groups by expanding access in providing a chance for *all* students instead of only a privileged few to participate in research and have ownership as co-authors on abstracts in contributing to new knowledge (Auchincloss et al., 2014). By focusing on regional fossils for research and across the course, students are able to synthesize and enrich their understanding of paleontological concepts by making connections in learning with their lives beyond school, not only following place-based approaches (Demarest, 2015) but modeling the methods and objectives promoted by culturally responsive pedagogy in desiring to broaden participation as well.

Self-Reflection and Action

The final piece, and perhaps the most important, as an initial step in shifting *your* mindset to prioritize a classroom in which all individuals are engaged and supported, is committing to personal reflection as an educator interested in broadening participation. The classroom that is inclusive and culturally responsive does not rely on a "checklist" of approaches. It is an ongoing practice that is intentional, and in order to be fully genuine in embracing the need for equity, objectives must come from within if you, as an instructor, are to honestly welcome and value diversity in the classroom. That requires not only a consideration of how to structure a course, but far more challenging work such as examining assumptions that you may have as an instructor that you bring to interactions with students. You must be willing to better understand who you are as a person, and how your identity is intersectional, influenced by your upbringing and work as a scientist among other aspects of your life, and how your values and beliefs affect how you operate as an educator. In essence, you must acknowledge your positionality, that is, how your perceptions and actions are shaped by your own experiences and that we are each unique in how we interact with the world. Everything we do from how we communicate to academic and social expectations is influenced by individual life experiences that are often strongly embedded in assumptions that derive from a cultural framework. It is important that you address biases you may have (e.g., do the Harvard Project Implicit Test), unlearn harmful behaviors, unpack any privilege, and assess how your perspectives are affected upon gaining this new knowledge about yourself. It is equally critical to maintain an awareness of power dynamics as well as practice mindfulness and empathy. You must work to recognize that where your students are coming from and where you are coming from are not necessarily aligned and that can be a roadblock to their success in the classroom and beyond. Teachers are often mismatched in being similar to their students in that they may be different in race/ethnicity or socioeconomic status (Villegas & Lucas, 2002; Kena et al., 2014; Milner, 2016), and in order to overcome such barriers that may inhibit connecting effectively with students, critical self-reflection by the instructor is essential. This personal work includes recognizing and combating issues such as "white gaze" or behaviors that are inherently embedded in ableism or sexism that may be deeply engrained in our culture. While we may not intend to harm others by our words or actions, if we're not aware of these issues, it impacts our ability to fully understand how we may view and experience the world quite differently compared to others and thereby limits our ability to effectively support *all* students (see Dutt, 2020, for more on these issues in geosciences). Furthermore, reflection is an ongoing process, and as you start to shift your practice to be more culturally responsive and inclusive, you will need to evaluate approaches that lead to success compared to those that are not 'failed' but rather opportunities to learn. These can include recognizing instances of cultural appropriation in the classroom (e.g., offensive use of culture as costumes) or engaging in behaviors that may in fact be problematic (such as using the idioms "hold down the fort" or "let's have a pow wow about that" as these phrases are insensitive to Indigenous populations). If you are not sure where to begin in this process of personal reflection as relevant in seeking to adopt culturally responsive pedagogy as your mindset and in your classroom, see Jackson (1993–1994) and Howard (2003) for more guidance and strategies for implementation.

Likewise, it is essential that you challenge yourself to actually effect significant change as part of pursuing action upon personal reflection. Adding a few cultural connections or modifying a lesson plan or your class policies to be more inclusive is a good start, but in order to fully cultivate a learning community (and ultimately a profession and more informed public) that features paleontology as a discipline that not only embraces but thrives on diversity, more work is needed. To that end, don't be afraid to get uncomfortable as part of learning. Don't shy away from difficulties in understanding the barriers and perspectives of people who may not be like you. Take the opportunity to educate yourself by attending workshops on culturally responsive pedagogy and meetings such as those held by the Society for the Advancement of Chicanos/Native Americans in Science (SACNAS), National Association of Black Geoscientists (NABG), or Geoscience Alliance Conference in support of broadening participation of Native Americans in Geoscience. Learn about marginalized and underserved groups and how best you can support them instead of relying on the labor and advocacy of people that belong to these groups (as they are already facing adversity in working to succeed compared to others that benefit from more privilege). Seek existing resources and read articles that address these issues to help in building a student-centered pedagogy that speaks to cultural competencies, sense of place, etc. Explore the rise in publications that focus on what we can do as educators in the classroom to support students from all backgrounds, particularly over the last few years out of the *Journal of Geoscience Education* (e.g., MacDonald et al., 2019; Núñez et al., 2020). Attend and present in conference sessions and webinars on broadening participation that have also increased in recent years due to growing collective action from within the paleontological community. Stay informed of opportunities and events by following organizations such as the International Association for Geoscience Diversity (IAGD), GeoLatinas, or other groups on social media that specifically promote and support initiatives related to diversity and education in paleontology and the geosciences overall. Finally, join and contribute to efforts in professional societies that work to improve equity, diversity, and inclusion through activities in support of recruitment and retention, including new policies related to nondiscrimination and issues of ethics (see the Code of Conduct by the Paleontological Society, 2019) to ensure a safe and inclusive community of practitioners in our field.*

Final Recommendations

To effectively broaden participation through improved pedagogy, consider the following question with respect to learners that vary in ability, background, and identity. Are your classroom approaches *accessible*, *engaging*, and *meaningful* for all individuals? To fully reach, support, and promote the success of all learners, it is critical that we, as a discipline, adopt more practices that are culturally responsive and recognize that such approaches are multifaceted and best executed within an integrated framework. So, how do we get started?

To best guide and inspire students in the classroom, not only do we need to understand who we are as the educator in guiding them on this journey and what we wish for them to learn about the fossil record, we need to continually recognize that our students are individuals who vary in their motivation, needs,

* During the months leading up to publication of this Element, worldwide protests against racism have led to new actions by individuals, organizations, and professional societies. More people are recognizing that they need to listen to people of color and acknowledge that discrimination is pervasive and change is needed, including in the geosciences. Numerous new materials for learning and action are now available as a result of these events, such as petitions, online publications, and web resources highlighting diverse geologists.

resources, etc. Working to learn about who they are is critical. Pre-surveys can be helpful for understanding the attitudes, skills, and knowledge that students have coming into a course, but they can also serve as an opportunity to gain insight as to who they are as individuals and what might be important to them. For example, initially, you might be interested to know what biology and geology courses students have completed or perhaps the level of understanding or any perceptions they may have about evolution or the utility of the fossil record overall. Learning about any experiences exploring in nature or going to museums could be useful, or if utilizing place-based approaches, where they have lived, visited, etc. Likewise, it could be interesting to know simply what matters to them as individuals in considering how aspects of paleontology could be connected to their lives. Finding a way to harness what motivates and excites diverse students in reflecting on how best to approach various subjects in the classroom can be extremely valuable and help to empower them to succeed. Drawing on models of culturally responsive pedagogy as well as a related need for growth as an instructor focused on broadening participation, here is a summary of my final recommendations:

- Become culturally literate and intentional in infusing cross-cultural perspectives in the classroom
- Build your knowledge of students as individuals and provide meaningful learning opportunities
- Develop a repertoire of materials and strategies that are accessible and engaging for all students
- Cultivate a sense of belonging for all individuals through collaborative and inclusive learning
- Consider your biases in self-reflection and how your experiences may influence your instruction and then work to effect the change that is needed

We have the power, and responsibility, as educators to make a difference in fighting the diversity crisis that exists in our discipline. Teaching in ways that recognize and incorporate the strengths of *all* individuals in the classroom is essential. By building on cultural contexts and places of meaning, as well as fostering inclusion and engaging a community of diverse learners, we can enhance access and equity, collectively advance scientific progress and broaden participation in our field, and in doing so improve the perception and value of paleontology overall. Note that any step, no matter how small, is a step in the right direction to becoming more culturally responsive and inclusive to better support the success of each and every single student in the classroom.

Appendix A

Demographic data for the Paleontological Society membership as of fall 2017 based on a 20 percent response rate from members (Yacobucci, pers. comm. March 2018)

GENDER	Man	64%
	Woman	32%
	Transgender	1%
	Prefer not to answer	3%
RACE/ ETHNICITY	White/Caucasian	76%
	Asian	7%
	Hispanic/Latino	7%
	Other	6%
	Black/African American	2%
	Pacific Islander	1%
	Prefer not to answer	1%
LGBTQ?	No	82%
	Yes	8%
	Prefer not to answer	9%
	Other	1%

Appendix B

Syllabus statement on equity, diversity, and inclusion drafted for use in my courses

I acknowledge that students have diverse backgrounds and intersectional identities. In addition, I am aware that students vary in their access and privilege to resources in support of their education. I am committed to improving equity in the classroom, welcoming diversity, and providing an inclusive community for learning that respects and meets the needs of all students. Traditionally, many voices have been marginalized or excluded in the sciences leading to underrepresentation. I am interested in broadening participation of underserved groups and recognize that diversity in the classroom enriches the experience for all. The contributions that unique perspectives bring to learning can better advance scientific progress and empower citizens through enhanced knowledge. It is my goal to utilize instructional approaches that are culturally responsive, honor all identities, and are inclusive in making learning accessible for everyone. Like many people, I am still learning about the challenges and biases that impact individuals daily. Please communicate any concerns you may have regarding the classroom atmosphere and environment for learning as needed. Your experience in this class is important; I look forward to understanding how best I can support and value students who vary in age, gender identity, race, ethnicity, language, neurotype, nationality, religion, disability, sexual orientation, socioeconomic status, and other aspects of identity and culture as part of guiding your participation in this course.*

* The author, upon further reflection following global protests to fight racism, is now working to revise the above statement to incorporate stronger anti-racist language and action in support of underrepresented groups in academia, geosciences, and overall.

References

Adams, M., Rodriguez, S. & Zimmer, K. 2017. *Culturally Relevant Teaching: Preparing Teachers to Include All Learners*. Rowman & Littlefield. Lanham, Maryland.

Aronson, B. & Laughter, J. 2016. The Theory and Practice of Culturally Relevant Education: A Synthesis of Research Across Content Areas. *Review of Educational Research* 86(1): 163–206.

Atchison, C. & Martinez-Frias, J. 2012. Inclusive Geoscience Instruction. *Nature Geoscience* 5: 366.

Auchincloss, L. C., Laursen, S. L., Branchaw, J. L. et al. 2014. Assessment of Course-Based Undergraduate Research Experiences: A Meeting Report. *CBE – Life Sciences Education* 13: 29–40.

Bearded Lady Project, 2018. Retrieved from: http://thebeardedladyproject.com/. Accessed January 15, 2018.

Bernard, R. E. & Cooperdock, E. H. G. 2018. No Progress on Diversity in 40 Years. *Nature Geoscience* 11: 292–295.

Black, R. Queer Voices in Paleontology. *Nature*. Careers Column. Retrieved from: https://www.nature.com/articles/d41586-019-02113-6. Accessed August, 9, 2019.

Brown-Jeffy, S. & Cooper, J. E. 2011. Toward a Conceptual Framework of Culturally Relevant Pedagogy: An Overview of the Conceptual and Theoretical Literature. *Teacher Education Quarterly* 38(1): 65–84.

Byrd, C. M. 2016. Does Culturally Relevant Teaching Work? An Examination from Student Perspectives. *SAGE Open* 6(3): 1–10.

Carabajal, I.G., Marshall, A.M., and Atchison, C.L. 2017. A Synthesis of Instructional Strategies in Geoscience Education Literature That Address Barriers to Inclusion for Students With Disabilities. *Journal of Geoscience Education* 65(4): 531–541.

CAST, 2018. Universal Design for Learning Guidelines version 2.2. Retrieved from: http://udlguidelines.cast.org. Accessed December 1, 2018.

Chávez, A. F. & Longerbeam, S. D. 2016. *Teaching Across Cultural Strengths*. Stylus Publishing. Sterling, VA.

Davies, C. P. 2006. Implementing Earth Systems Science Curriculum: Evaluating the Integration of Urban Environments for an Urban Audience. *Journal of Geoscience Education* 54(3): 364–373.

De Paor, D., Karabinos, P., Dickens, G. & Atchison, C. 2017. Color Vision Deficiency and the Geosciences. *GSA Today* 27(6): 42–43.

Demarest, A. B. 2015. *Place-based Curriculum Design*. Routledge. Taylor Francis Group. New York, NY.

Duffin, C. J. 2008. Fossils as Drugs: Pharmaceutical Palaeontology. *Ferrantia* 54: 1–83.

Dutt, K. 2020. Race and Racism in the Geosciences. *Nature Geosciences* (13): 2–3.

Gay, G. 2000. *Culturally Responsive Teaching: Theory, Research, and Practice*. Teachers College Press. Columbia University. New York.

Gay, G. 2018. *Culturally Responsive Teaching: Theory, Research, and Practice*. 3rd ed. Multicultural Education Series. J.A. Banks, ed. Teachers College Press. Columbia University. New York.

Geraghty Ward, E. M., Semken, S. & Libarkin, J. C. 2014. The Design of Place-Based, Culturally Informed Geoscience Assessment. *Journal of Geoscience Education* 62: 86–103.

Gold, M. E. L. & West, A. R. 2017. *She Found Fossils*. CreateSpace Independent Publishing Platform.

Goldberg, E. 2019. Earth Science Has a Whiteness Problem. New York Times.

Gonzales, P. M., Blanton, H. & Williams, K. J. 2002. The Effects of Stereotype Threat and Double-Minority Status on the Test Performance of Latino Women. *Personality and Social Psychology Bulletin* 28: 659–670.

Gore, P. J. W. & Witherspoon, W. 2013. *Roadside Geology of Georgia*. Mountain Press. Missoula, Montana.

Goring, S., Whitney, K. S., Jacob, A., Bruna, E. & Poisot, T. 2017. Making Scientific Content More Accessible. *Authorea*. Retrieved from: https://doi.org/10.22541/au.150844289.92609826. Accessed November 2, 2018.

Gruenewald, D. A. and Smith, G.A., eds. 2014. Place-Based Education in the Global Age: Local Diversity. Psychology Press. Taylor & Francis Group. New York and London.

Guterl, F. 2014. The Inclusion Equation. *Scientific American* 311(4): 38–40.

Hailu, M.F., Pan, J., Mackey, J.Z., and Arend, B. 2017. Turning Good Intentions into Good Teaching: Five Common Principles for Culturally Responsive Pedagogy. In *Promoting Intercultural Communication Competencies in Higher Education* (pp. 20–53). IGI Global.

Hammond, Z. 2015. *Culturally Responsive Teaching and the Brain*. Corwin. Thousand Oaks, CA.

Heise, U. K. 2016. *Imagining Extinction*. The University of Chicago Press. Chicago, IL.

Hewitt, C. 2017. Pan-African: Introduction to a Pedagogical Approach. Retrieved from: https://serc.carleton.edu/integrate/workshops/african-education/essay/180254.html. Accessed January 15, 2018.

Holmes, M. A., O'Connell, S., Frey, C. & Ongley, L. 2008. Gender Imbalance in U.S. Geoscience Academia. *Nature Geoscience* 1(2): 79–82.

Howard, T. C. 2003. Culturally Relevant Pedagogy: Ingredients for Critical Teacher Reflection. *Theory into Practice*: 42(3): 195–202 – Teacher Reflection and Race in Cultural Contexts.

Howard Hughes Medical Institute. 2013. BioInteractive Video: https://www.biointeractive.org/classroom-resources/making-fittest-got-lactase-coevolution-genes-and-culture. Accessed January 15, 2018.

Hughes, B.E. 2018. Coming out in STEM: Factors affecting retention of sexual minority STEM students. *Science Advances* 4(3): eaao6373.

Hughes, N., Ghosh, P. & Bhattacharya, D. 2015. The Monishar Pathorer Bon (Monisha and the Stone Forest) Book Project: Novel Educational Outreach to Children in Rural Communities, Eastern Indian Subcontinent. *Journal of Geoscience Education* 63(1): 18–28.

Huntoon, J. E. & Lane, M. 2007. Diversity in the Geosciences and Successful Strategies for Increasing Diversity. *Journal of Geoscience Education* 55(6): 447–457.

Huntoon, J.E., Tanenbaum, C., and Hodges, J. 2015. Increasing Diversity in the Geosciences. *Eos* 96. doi:10.1029/2015EO025897.

Inclusive Astronomy. 2015. Endorsed by the American Astronomical Society Council. Retrieved from: https://tiki.aas.org/tiki-index.php?page=Inclusive_Astronomy_The_Nashville_Recommendations. Accessed December 1, 2018.

Irvin, J. L. & Darling, D. 2005. What Research Says: Improving Minority Student Achievement by Making Cultural Connections. *Middle School Journal* 36(5): 46–50.

Jackson, F. R. 1993–1994. Seven Strategies to Support a Culturally Responsive Pedagogy. *Journal of Reading* 37(4): 298–303.

Jacobs, L. L. 1993. *Quest for the African Dinosaurs*. New York: Villard Books.

Jimenez, M.F., Laverty, T.M., Bombaci, S.P., Wilkins, K., Bennett, D.E., and Pejchar, L. 2019. Underrepresented faculty play a disproportionate role in advancing diversity and inclusion. Nature Ecology & Evolution 3(7): 1030–1033.

Kena, G., Aud, S., Johnson, F., Wang, X., Zhang, J., Rathbun, A., Wildinson-Fliker, S., and Kristapovich, P. 2014. *The Condition of Education*. Washington, DC: U.S. Department of Education, National Center for Education Statistics. 256 pp.

Kenworthy, J. & Santucci, V. 2006. A Preliminary Inventory of National Park Service Paleontological Resources in Cultural Resource Contexts, Part 1: General Overview. In *Fossils from Federal Lands, New Mexico Museum of Natural History and Science Bulletin* 34: 70–76.

Kirkby, K. C. 2014. Place in the City: Place-based Learning in a Large Urban Undergraduate Geoscience Program. *Journal of Geoscience Education* 62: 177–186.

Ladson-Billings, G. 1995. But That's Just Good Teaching! The Case for Culturally Relevant Pedagogy. *Theory into Practice* 34(3): *Culturally Relevant Teaching*: 159–165.

Ladson-Billings, G. 2014. Culturally Relevant Pedagogy 2.0: A.k.a. the Remix. *Harvard Educational Review* 84(1): 74–84.

MacDonald, R.H., Beane, R.J., Baer, E.M., Eddy, P.L., Emerson, N.R., Hodder, J., Iverson, E.R., McDaris, J.R., O'Connell, K., Orman, C.J. 2019. Accelerating change: The power of faculty change agents to promote diversity and inclusive teaching practices. Journal of Geoscience Education 67(4): 330–339.

Marino, E. 2015. *Fierce Climate, Sacred Ground: An Ethnography of Climate Change in Shishmaref, Alaska*. University of Alaska Press. Fairbanks, AK.

Mayor, A. 2000. *The First Fossil Hunters*. Princeton University Press. Princeton, NJ.

Mayor, A. 2005. *Fossil Legends of the First Americans*. Princeton University Press. Princeton, NJ.

Mayor, A. 2007a. Fossils in Native American Lands – Whose Bones, Whose Story? Fossil Appropriations Past and Present. *History of Society Annual Meeting Paper*.

Mayor, A. 2007b. Place Names Describing Fossils in Oral Traditions. *Geological Society London Special Publications* 273(1): 245–261.

McCartney, T. L. & Lee, T. S. 2014. Critical Culturally Sustaining/Revitalizing Pedagogy and Indigenous Education Sovereignty. *Harvard Educational Review* 84(1): 101–124.

Medin, D. L. & Lee, C. D. 2012. Diversity Makes Better Science. *Observer*: 25(5).

Meyer, X. S. & Crawford, B. A. 2011. Teaching Science As a Cultural Way of Knowing: Merging Authentic Inquiry, Nature of Science and Multicultural Strategies. *Cultural Studies of Science Education* 6: 525–547.

Meyer, X. S., Capps, D. K., Crawford, B. A. & Ross, R. 2012. Using Inquiry and Tenets of Multicultural Education to Engage Latino English-Language Learning Students in Learning About Geology and the Nature of Science. *Journal of Geoscience Education* 60: 212–219.

Milner, H.R. 2016. A Black Male Teacher's Culturally Responsive Practices. *The Journal of Negro Education* 85(4): 417–432.

Núñez, A.M., Rivera, J., and Hallmark, T. 2020. Applying an intersectionality lens to expand equity in the geosciences, Journal of Geoscience Education, 68(2): 97–114.

O'Connell, S. & Holmes, M. A. 2011. Obstacles to the Recruitment of Minorities into the Geosciences: A Call to Action. *GSA Today* 21(6): 52–54.

Paleontological Society. 2019. Paleontological Society Policy and Non-Discrimination and Code of Conduct. Retrieved from: www.paleosoc.org/paleontological-society-policy-on-non-discrimination-and-code-of-conduct/. Accessed August 9, 2019.

Paris, D. 2012. Culturally Sustaining Pedagogy: A Needed Change in Stance, Terminology, and Practice. *Educational Researcher* 41(3): 93–97.

Paris, D. & Alim, H. S. 2014. What Are We Seeking to Sustain Through Culturally Sustaining Pedagogy? A Loving Critique Forward. *Harvard Educational Review* 84(1): 85–100.

Patterson, J. L., Conolly, M. C. & Ritter, S. A. 2009. Restructuring the Inclusion Classroom to Facilitate Differentiated Instruction. *Middle School Journal* 41(1): 46–52.

Pewewardy, C. 1993. Culturally Responsible Pedagogy in Action: An American Indian Magnet School. In E. Hollins, J. King & W. Hayman, eds. *Teaching Diverse Populations: Formulating a Knowledge Base*. Albany, NY: State University of New York Press, pp. 77–92.

Phillips, K. 2014. How Diversity Works. *Scientific American* 311(4): 42–47.

Plotnick, R., Stigall, A. L. & Stefanescu, I. 2014. Evolution of Paleontology: Long-Term Gender Trends in an Earth Science Discipline. *GSA Today* 24(11): 44–45.

Powell, K. 2018. These Labs Are Remarkably Diverse – Here's Why They're Winning at Science. *Nature* 558: 19–22.

Rull, V. & Giralt, S. 2018. Editorial: Paleoecology of Easter Island: Natural and Anthropogenic Drivers of Ecological Change. *Frontiers in Ecology and Evolution* 6: 105.

Santamaria, L. J. 2009. Culturally Responsive Differentiated Instruction: Narrowing Gaps Between Best Pedagogical Practices Benefiting All Learners. *Teachers College Record* 111(1): 214–247.

Semken, S. 2005. Sense of Place and Place-Based Introductory Geoscience Teaching for American Indian and Alaska Native Undergraduates. *Journal of Geoscience Education* 53 (2): 149–157.

Semken, S. & Butler Freeman, C. 2008. Sense of Place in the Practice and Assessment of Place-Based Science Teaching. *Science Education* 92(6): 1042–1057.

Semken, S., Butler Freeman, C., Watts, N. B. et al. 2009. Factors That Influence Sense of Place As a Learning Outcome and Assessment Measure of Place-Based Geoscience Teaching. *Electronic Journal of Science Education* 13(2): 1–25.

Senos, R., Lake, F., Turner, N. & Martinez, D. 2006. Traditional Ecological Knowledge and Restoration Practice. In *Restoring the Pacific Northwest: The Art and Science of Ecological Restoration in Cascadia*, D. Apostol & M. Sinclair, eds. Washington, WA: Island Press, pp. 393–426.

Sexton, J.M., Pugh, K.J., Bergstrom, C.M., and Riggs, E.M. 2018. Reasons undergraduate students majored in geology across six universities. *Journal of Geoscience Education* 66(4): 319–336.

Seymour, E. & Hewitt, N. M. 1997. *Talk about Leaving: Why Undergraduates Leave the Sciences*. Boulder, CO: Westview Press.

Sherman-Morris, K. & McNeal, K. S. 2016. Understanding Perceptions of the Geosciences Among Minority and Nonminority Undergraduate Students. *Journal of Geoscience Education* 64: 147–156.

Shubin, N. 2008. *Your Inner Fish*. Pantheon Books. New York, NY.

Sleeter, C. E. 2011. An Agenda to Strengthen Culturally Responsive Pedagogy. *English Teaching: Practice and Critique* 10(2): 7–23.

Sleeter, C. E. 2012. Confronting the Marginalization of Culturally Responsive Pedagogy. *Urban Education* 47(3): 562–584.

Smith-Doerr, L., Alegria, S. N. & Sacco, T. 2017. How Diversity Matters in The U.S. Science and Engineering Workforce: A Critical Review Considering Integration in Teams, Fields, and Organizational Contexts. *Engaging Science, Technology, and Society* 3: 139–153.

Sobel, D. 2004. *Place-Based Education. Connecting Classrooms and Communities*. The Orion Society. Great Barrington, MA.

Sodikoff, G. M., ed. 2012. *The Anthropology of Extinction. Essays on Culture and Species Death*. Indiana University Press. Bloomington, IN.

Steele, C. M. 1997. A Threat in the Air: How Stereotypes Shape Intellectual Identity and Performance. *American Psychologist* 52(6): 613–629.

Stigall, A. L. 2013. Where Are the Women in Paleontology? *Priscum* 20(1): 1–3.

Stokes, P. J., Levine, R. & Flessa, K. W. 2014. Why Are There So Few Hispanic Students in Geoscience? *GSA Today* 24(1): 52–53.

Stokes, P.J., Levine, R. & Flessa, K.W. 2015. Choosing the Geoscience Major: Important Factors, Race/Ethnicity, and Gender. *Journal of Geoscience Education* 63: 250–263.

Stricker, B. 2017. *Daring to Dig*. Paleontological Research Institution Special Publications. Paleontological Research Institution. Ithaca, NY.

Velasco, A. A. & Jaurrieta de Velasco, E. 2010. Striving to Diversify the Geosciences Workforce. *Eos* 91(33): 289–290.

Villegas, A. M. & Lucas, T. 2002. Preparing Culturally Responsive Teachers. Rethinking the Curriculum. *Journal of Teacher Education* 53(1): 20–32.

Williams, B. A. & Lemons-Smith, S. 2009. Perspectives on Equity and Access in Mathematics and Science for a 21st-Century Democracy: Re-Visioning Our Gaze. *Democracy & Education* 18(3): 23–28.

Wilson, L. S. 1992. The Benefits of Diversity in the Science and Engineering Work Force. Chapter 1 in *Science and Engineering Programs: On Target for Women?* The National Academies Press, Washington, DC. pp. 1–14.

Wlodkowski, R. J. & Ginsberg, M. B. 1995. A Framework for Culturally Responsive Teaching. *Strengthening Student Engagement* 53(1): 17–21.

Yusoff, K. 2019. *A Billion Black Anthropocenes or None*. University of Minnesota Press, Minneapolis, MN.

Acknowledgments

Thank you to Rowan Lockwood, Phoebe Cohen, and Lisa Park Boush for embracing the idea for this work as part of the 2018 short course and for support along the way in making it a reality including the funding provided by the Paleontological Society to present at GSA. Dena Smith, I am indebted to you for your perspectives and guidance through my early work on this manuscript and overall. Thanks also to Margaret Fraiser for motivating conversations and learning opportunities, as well as a growing and inspiring community of colleagues who are passionate about diversity and inclusion in paleontology. I am extremely grateful to have attended and gained insight as to the importance of intersections between culture and pedagogy in broadening participation from the Teaching Geosciences with Pan-African Approaches workshop organized by SERC. Thanks to anonymous reviewers for helping to improve this manuscript, Rebecca Barrett-Fox and the AGT writing challenge for encouraging my success, and to my former student, Jessica Martinez, for being a remarkable person to work with on science education projects. In addition, I am incredibly appreciative of my family who have steadfastly supported me through every step of this journey. Finally, special gratitude is extended to my many wonderful students at Georgia State University; I am a much better instructor because of you.

Elements of Paleontology

Editor-in-Chief
Colin D. Sumrall
University of Tennessee

About the Series

The Elements of Paleontology series is a publishing collaboration between the Paleontological Society and Cambridge University Press. The series covers the full spectrum of topics in paleontology and paleobiology, and related topics in the earth and life sciences of interest to students and researchers of paleontology.

The Paleontological Society is an international nonprofit organization devoted exclusively to the science of paleontology: invertebrate and vertebrate paleontology, micropaleontology, and paleobotany. The Society's mission is to advance the study of the fossil record through scientific research, education, and advocacy. Its vision is to be a leading global advocate for understanding life's history and evolution. The Society has several membership categories, including regular, amateur/avocational, student, and retired. Members, representing some forty countries, include professional paleontologists, academicians, science editors, earth science teachers, museum specialists, undergraduate and graduate students, postdoctoral scholars, and amateur/avocational paleontologists.

Cambridge Elements

Elements of Paleontology

Elements in the Series

These Elements are contributions to the Paleontological Short Course on *Pedagogy and Technology in the Modern Paleontology Classroom* (organized by Phoebe Cohen, Rowan Lockwood and Lisa Boush), convened at the Geological Society of America Annual Meeting in November 2018 (Indianapolis, Indiana USA).

Flipping the Paleontology Classroom: Benefits, Challenges, and Strategies
Matthew E. Clapham

Integrating Macrostrat and Rockd into Undergraduate Earth Science Teaching
Pheobe A. Cohen, Rowan Lockwood, and Shanan Peters

Student-Centered Teaching in Paleontology and Geoscience Classrooms
Robyn Mieko Dahl

Beyond Hands On: Incorporating Kinesthetic Learning in an Undergraduate Paleontology Class
David W. Goldsmith

Incorporating Research into Undergraduate Paleontology Courses: Or a Tale of 23,276
Mulinia Patricia H. Kelley

Utilizing the Paleobiology Database to Provide Educational Opportunities for Undergraduates
Rowan Lockwood, Pheobe A. Cohen, Mark D. Uhen, and Katherine Ryker

Integrating Active Learning into Paleontology Classes
Alison N. Olcott

Dinosaurs: A Catalyst for Critical Thought
Darrin Pagnac

Confronting Prior Conceptions in Paleontology Courses
Margaret M. Yacobucci

The Neotoma Paleoecology Database: A Research Outreach Nexus
Simon J. Goring, Russell Graham, Shane Oeffler, Amy Myrbo, James S. Oliver, Carol Ormond, and John W. Williams

Equity, Culture, and Place in Teaching Paleontology: Student-Centered Pedagogy for Broadening Participation
Christy C. Visaggi

A full series listing is available at: www.cambridge.org/EPLY

CPSIA information can be obtained
at www.ICGtesting.com
Printed in the USA
LVHW050957210820
663744LV00013B/411